IMAGES
of America

THE US ARMY CORPS OF ENGINEERS ON THE MISSISSIPPI RIVER

On the Cover: Workers help clear and grade the bank of the Mississippi River near Memphis, Tennessee, to prepare for the installation of a revetment, or bank protection. (Courtesy of the US Army Corps of Engineers, Memphis District.)

IMAGES
of America

The US Army Corps of Engineers on the Mississippi River

Damon Manders

ISBN 978-1-4671-0860-7

Published by Arcadia Publishing
Charleston, South Carolina

Printed in the United States of America

Library of Congress Control Number: 2022935539

For all general information, please contact Arcadia Publishing:
Telephone 843-853-2070
Fax 843-853-0044
E-mail sales@arcadiapublishing.com
For customer service and orders:
Toll-Free 1-888-313-2665

Visit us on the Internet at www.arcadiapublishing.com

Dedicated to the employees of the US Army Corps of Engineers, whose work continues to contribute to the safety and prosperity of the Mississippi Valley.

Contents

Acknowledgments

The US Army Corps of Engineers has one of the most robust historical programs in the federal government. Most of its projects, however, focus primarily on illustrated textual histories. When given the opportunity to write a history that leveraged the wealth of Corps of Engineers photographs to tell the story of engineering on the Mississippi River, I was excited to participate. The number of images the Corps has collected is immense, covering more than a century of projects and documenting every aspect of its work. While some of these images have been used in the past, I am pleased to introduce many others discovered during my research. Unless otherwise noted, all images in this book are used courtesy of the US Army Corps of Engineers, Mississippi Valley Division.

I wish to express my thanks to the many individuals across the US Army Corps of Engineers who helped me in the procurement of these photographs, without whom I would not have been able to assemble this book. These include Mississippi Valley Division historian Brian Rentfro, the Memphis District public affairs office, and New Orleans District photographer Nancy Mayberry. I wish to extend my appreciation to the St. Louis Art Museum, Louisiana State Archives, and the St. Louis Mercantile Library for their assistance in tracking down additional images. In addition, I am grateful for the textual review provided by Mr. Rentfro. I also thank my wife and children for putting up with the long hours I spent outside of my nine-to-five job to complete this book. Most of all, I wish to thank God for the opportunity to work on this project.

Introduction

From the earliest settlements on the Mississippi River, the primary challenge for people in the valley has been how to manage the dangers of living near the river while extracting its benefits. The river provided fresh water, food, rich soils, and easy transportation, but flooding, bank erosion, and shoaling made living by or using the river dangerous at times. Native Americans and early traders could abandon seasonal dwellings during floods and avoid unnavigable stretches of the river. With the establishment of permanent settlements, the challenges of the river became more difficult. Engineers began to make flood control and navigational improvements, which ultimately became the responsibility of the US Army Corps of Engineers.

For most of the early history of the valley, individual landowners were responsible for building protection, such as levees, dikes, or drainage canals. Only in cities such as New Orleans, Louisiana, were construction efforts well organized and engineered. The establishment of states in the valley after 1820 enabled governments to make more robust efforts by placing police juries over flood control, establishing state engineering offices, or creating local levee districts.

At first, federal involvement was limited to exploration and surveys. Prior to 1824, many in Congress believed navigation and flood control to be local matters. Once the Supreme Court ruled that constitutional authority to regulate commerce included maintaining navigable waterways, the government gave this responsibility to the Corps of Engineers. As continued flooding revealed weaknesses in levees, Congress authorized hydrological surveys to recommend methods to control floods. After the Civil War, the Corps became more involved in river and harbor improvements, including regulation of local activities. With the increasing success of these efforts, local governments lobbied Congress to take coordinated action.

In 1879, Congress created the Mississippi River Commission to improve river navigation. Initially, the commission focused on a plan of improvement that used channel contraction in low water to speed the river and scour the bottom while protecting the banks from erosion. The commission accepted that a slower river dropped more sediment and contributed to sandbars, but it debated the best means of addressing the issue. Unfortunately, constant damages to works from flooding and inconsistent funding prevented much more than experimentation until another flood in 1882 resulted in Congress placing Corps of Engineers personnel under commission control in four districts.

In 1880, the commission approved the construction of levees to constrict the river and help scour the riverbed during high water. Congress afterward approved repair of levees, but always under the guise of navigational improvements—it did not overtly approve flood control until 1917. Over time, the commission proposed, and Congress accepted, the construction of levees as the most effective and enduring improvement. This was the beginning of the so-called "levees only" policy, although the commission had always made other improvements as well.

The Great Flood of 1927 was one of the most devastating disasters the nation has ever experienced. While other disasters caused more loss of life or property damage, the 1927 flood impacted the largest geographic area—a stretch 100 to 150 miles wide from Missouri to below New Orleans. The flood caused more than $12 billion in damages in current dollars and killed between 200 and 500 people. A combination of heavy rainfall and winter ice melt led to record river stages. The flood volume—2.3 million cubic feet per second at Arkansas City, Arkansas—became the volume used in future planning. This high water caused breaks in the levees from Arkansas to Louisiana from March to May. The total number of crevasses was between 120 and 225 on both federal and local levees, depending on who counted them (Manders, *Common Stock* 23–32).

In response to these floods, the federal government made the greatest rescue and recovery efforts to that date. While private donations and the American Red Cross delivered most assistance, the government provided logistical support. Pres. Calvin Coolidge assigned then–secretary of commerce Herbert Hoover to head these efforts. Aiding him was the chief of engineers, Maj.

Gen. Edgar Jadwin. The Mississippi River Commission helped with flood fighting and rescue efforts. With rising flood stages in late April, New Orleans and Louisiana state leaders requested that the commission and Corps of Engineers approve dynamiting the Caernarvon levee south of the city to lower flood stages. This proved once and for all that levees alone could not protect the valley, leading to a change in policy.

Within months of the 1927 flood, the Mississippi River Commission submitted a comprehensive plan for protecting the Mississippi Valley, but General Jadwin deemed it too expensive. Instead, he submitted his own plan, which the president sent to Congress. It included one controlled spillway at Bonnet Carré, Louisiana; three uncontrolled floodways using "safety plug" levees on the Atchafalaya River, the Boeuf Basin, and from Birds Point to New Madrid, Missouri; and higher levees, for less than $300 million. The safety-plug levees were shorter levees that would overtop on reaching a certain height without a spillway. The Jadwin Plan became the basis of the Mississippi River and Tributaries (MR&T) Project approved in the Flood Control Act of 1928 (Camillo and Pearcy 149–164; Manders, *Common Stock* 45–68).

Jadwin and the new president of the Mississippi River Commission, Brig. Gen. Thomas Jackson, started construction of the Bonnet Carré Spillway. The 1928 act placed the commission directly under the Corps of Engineers. It immediately ran into legal issues in forcing the purchase of land rights that delayed construction of the three floodways. In 1932, the new chief of engineers, Maj. Gen. Lytle Brown, and commission president Brig. Gen. Harley B. Ferguson implemented an experimental program of dredging and cutoffs. Over a dozen years, 16 cutoffs shortened the river by nearly 500 miles, greatly speeding floods to the gulf and reducing flood heights. With this and the addition of reservoirs built on tributaries, Ferguson was able to eliminate the floodway at Boeuf and recommend a smaller controlled floodway at Morganza, Louisiana.

The Bonnet Carré Spillway came into operation for the first time in 1937. Despite being incomplete, the Birds Point–New Madrid Floodway also came into operation. The Corps completed the construction of the Morganza Floodway in 1954 (Camillo and Pearcy 207–247; Manders, *Common Stock* 129–198). To address problems with the Atchafalaya River capturing an increased percentage of the Mississippi River, the commission also added a control structure at Old River, Louisiana, completed in 1962, to maintain a consistent 30-percent diversion (Reuss 207–248). In 1957, the commission completed a study of the project flood that resulted in the final major changes in the plan until 1973 (Manders, *Common Stock* 213–222).

The period from 1951 until 1965 was relatively flood-free, allowing the Corps of Engineers to make significant progress in the MR&T Project. The project was mostly fixed by that time, with only minor adjustments in levees or the addition of local pump facilities or navigation canals. With the advent of the environmental movement, there was growing concern about the environmental impact flood control projects had on the river. The Corps increasingly considered issues such as the protection of endangered species and restoration of gradually disappearing wetlands. By the 1990s, this included attempts to restore infusions of river water and silt into wetlands through freshwater diversions.

Several major floods tested the MR&T Project. The flood of 1973 damaged the Old River Control Structure. The flood of 1993 caused extensive damage to the Middle River. There were, in addition, several major hurricanes that threatened the system in Louisiana, from Hurricane Betsy in 1965 to Hurricane Katrina in 2005. Perhaps the most important test came in 2011, the first year the Corps of Engineers put the entire flood control system into operation and the first year it handled a flood approaching the size of the 1927 flood without a major breach in the federal levees.

Challenging floods would continue, but the Corps of Engineers demonstrated it could reduce the risk to life and property through structures it has built, although continued development of the Mississippi Valley has made use of the system more and more costly. The greater challenge remained how to balance those needs with protecting nature. While some have argued that the MR&T system is irreparably harming nature, others believe that lives must be protected at any cost. What people of the valley have made repeatedly clear over the centuries is that it is possible to find engineering solutions to whatever problems they face.

One

EARLY RIVER ENGINEERING

1820–1879

The first steamboat arrived in St. Louis, Missouri, in 1817. By the 1830s, there were thousands of steamboats docking each year (Manders and Rentfro viii–ix). This painting, *St. Louis in 1846*, by Henry Lewis, shows the St. Louis waterfront from the Illinois shore. (Courtesy of the Saint Louis Art Museum, Eliza McMillan Trust.)

In 1820, Maj. Joseph Totten (pictured) and Brig. Gen. Simon Bernard made the first survey of the Mississippi River. Soon after, Congress passed the 1824 General Survey Act and 1826 Rivers and Harbors Act, authorizing the first experiments with dikes, removal of obstacles, and other works on the Mississippi.

In 1826, Pres. John Quincy Adams selected Henry M. Shreve (pictured) as superintendent of western river improvement, with responsibilities to improve navigation on the Mississippi River. A former steamship developer, businessman, and blockade runner during the War of 1812, Shreve developed a steam vessel designed to remove snags (dead fallen trees).

Similar to this snag boat depicted in a sketch from *Harper's Weekly*, Shreve's snag boat, the *Heliopolis*, was a shallow-draft steamboat with a crane on the split foredeck capable of lifting even the largest trees (Clay 8). (Courtesy of the US Army Corps of Engineers, Memphis District.)

A major obstacle that Shreve removed was the Red River Raft—a large mass of fallen trees and driftwood that blocked 160 miles of this tributary of the Mississippi at multiple bends. This photograph from 1870 shows the raft as it then appeared; it reformed several times (Reuss 25–36). (Courtesy of the State Library of Louisiana.)

Although manual labor was also necessary, Shreve's snag boat was instrumental in removing obstacles throughout the Mississippi Valley, greatly reducing accidents. The Corps of Engineers replicated its work on other river systems nationwide.

The chief means of flood protection on the Mississippi River was through levees, built initially by private landowners and later by state and local levee organizations. In the 19th century, levee construction remained mostly a manual process using horse and wheelbarrow, similar to the scene in this undated photograph.

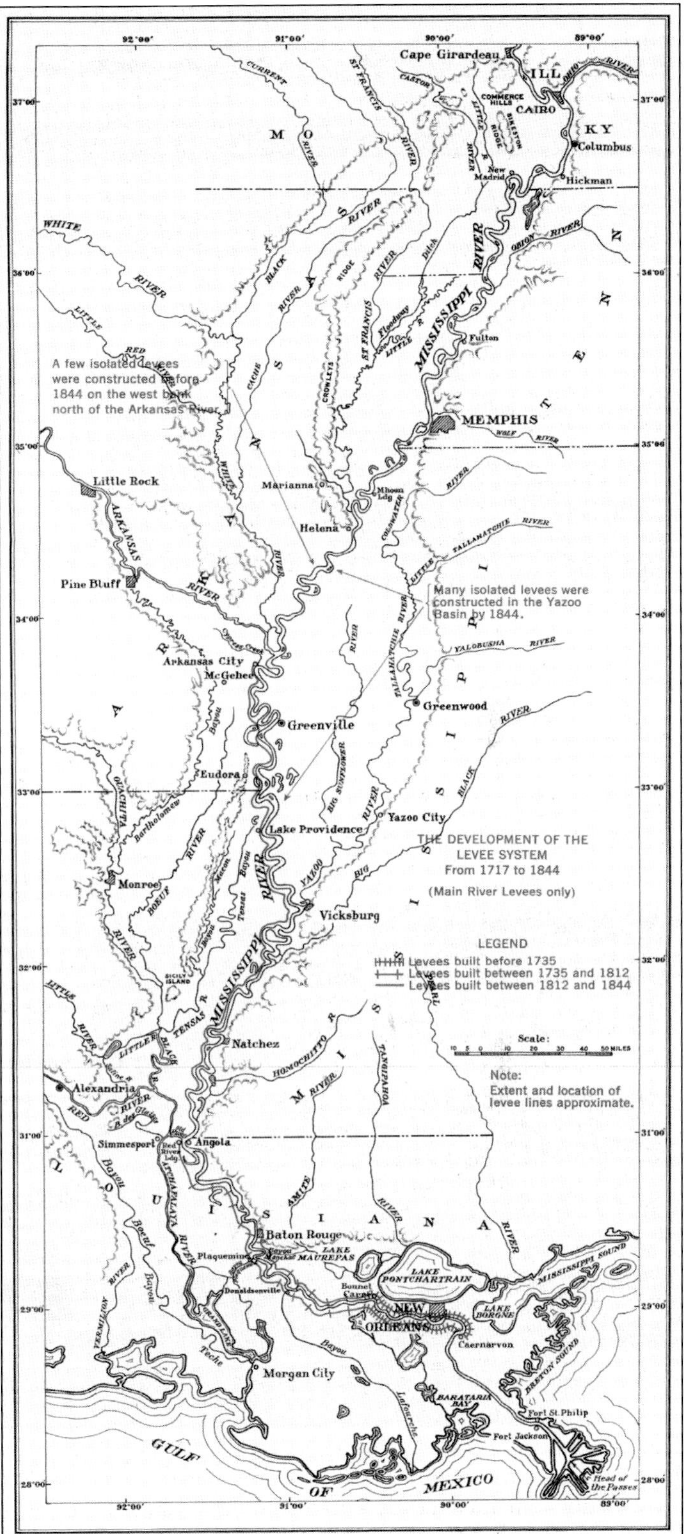

This map shows the development of the Mississippi River levee system from 1716 to 1844. By the mid-19th century, there were lines of levees for many miles, especially near major cities. However, the levees were inconsistent in their quality of engineering and materials.

Following a major flood in 1850, Congress authorized the first major hydrographic study of the Mississippi River. The War Department first hired civilian engineer Charles Ellet Jr. Although very general in scope, his report of 1852 recommended a combination of flood control works later included in the Mississippi River and Tributaries Project, including reservoirs to contain floodwaters upstream for later release.

The second hydrographic study of the Mississippi River funded by the War Department was that of Capt. Andrew A. Humphreys in 1861. One of the most scientific works of the century, its recommendation of using mainly levees to protect the Mississippi Valley became authoritative. Largely based on this work, Humphreys went on to serve as the chief of engineers from 1866 to 1879.

Lt. Henry L. Abbot assisted Humphreys in completing his survey of the Mississippi River. He went on to be a prominent instructor and researcher at the US Army Engineer School at Willets Point, New York.

An 1875 commission headed by Brevet Maj. Gen. Gouverneur K. Warren largely confirmed the findings of Humphreys and Abbot that levees provided the best protection against flooding, but it also recommended greater federal involvement in protecting the Mississippi Valley. (Courtesy of the Library of Congress.)

Floods were frightful experiences. This c. 1870 newspaper sketch by artist Charles Upham shows refugees being rescued.

River towns expanded rapidly due to increasing commerce on the Mississippi River. Cairo, Illinois, was a growing industrial center by the end of the 19th century, and was expected to become a major metropolis in the 20th.

Perhaps the most famous civil engineer of the post–Civil War era was James B. Eads, a self-educated engineer who invented an early diving bell and learned river mechanics as a salvager. He often came into conflict with the Corps of Engineers.

Eads built and designed the St. Louis Bridge, which was the longest arch bridge in the world at the time of its completion. This c. 1870 photograph shows the bridge under construction (Manders and Rentfro 23–28).

This photograph shows the Eads Bridge in St. Louis after its completion in 1874. It was considered a marvel in engineering (Manders and Rentfro 23–28).

Prior to 1876, the three heavily sedimented passes leading from the Mississippi River to the Gulf of Mexico were almost impassible due to their shallow depth. Ocean-going vessels often had to unload and reload cargo onto riverboats (Camillo and Pearcy 1–15). This aerial photograph of the South Pass dates from 1933. (Courtesy of the US Army Corps of Engineers, New Orleans District.)

In 1877, James Eads opened the South Pass of the river to traffic by using jetties—partly submerged dikes that channeled and sped the water—to scour the bottom. This photograph shows the jetties being repaired in the 1940s. (Courtesy of the US Army Corps of Engineers, New Orleans District.)

By the 1870s, river traffic was beginning to grow rapidly. This lithograph of Memphis Harbor dates from 1870 (Clay 28–29). (Courtesy of the US Army Corps of Engineers, Memphis District.)

This lithograph by St. Louis artist Charles Overall shows the St. Louis waterfront from the Eads Bridge in 1875 (Manders and Rentfro 40).

Two

The Mississippi River Commission

1879–1927

Congress established the Mississippi River Commission in 1879. Initial members included civil engineers James B. Eads of Missouri and Benjamin M. Harrod of Louisiana; politician (later US president) Benjamin Harrison; Coast and Geodetic Survey member Henry Mitchell; and US Army Corps of Engineers officers Brevet Brig. Gens. Charles R. Suter, Cyrus B. Comstock, and Quincy A. Gillmore, who was commission president (Camillo and Pearcy 38–42).

One of the first areas the commission addressed was the sandbar-choked Plum Point Reach between Hales Point, Tennessee, and Osceola, Arkansas. The commission used dikes to channel the river and revetment or groins to protect the banks. Pictured here is the state of the dikes in 1895, ten years after completion.

This unnamed dike in 1880 shows how the Plum Point dike appeared under construction. The commission initially made the dikes out of wood, which was lower cost but also more susceptible to flood damage.

This 1923 photograph shows the close weave of a dike, which trapped sediment to help build up banks and improve the channel. The commission experimented widely with construction methods and materials.

The commission later built numerous other dikes, such as this one at Sulphur Springs, Missouri, shown in 1901. Although construction materials have changed, dikes remain an effective means of shaping the channel, reducing the need for seasonal dredging.

Dikes were especially effective in the northern part of the river, which was narrower and less sedimented, as with this dike at Curtis Bend, Iowa.

Depicted here is a groin near Natchez, Mississippi. A groin was a wooden structure similar to a dike used to protect a bank, trap sand, and prevent meandering or bank caving.

The commission recommended the construction of levees as a secondary means of improving the channel during high water. As shown in this undated photograph, local levee construction was a labor-intensive and expensive process involving many people and teams of animals.

Until the 20th century, local levee districts relied primarily on manual labor—mostly convicts or the poor and disenfranchised—to build levees by hand using shovels and wheelbarrows.

Even into the 20th century, access still limited construction to manual earth-moving techniques at many points on the river. As labor costs increased, other methods became more cost-effective.

By the 20th century, steam or gasoline-powered vehicles were more common, like these tractors used by the St. Francis Levee District at Wyanoke, Arkansas, in 1922.

Using cranes or draglines to help move earth also become more common by the 1920s. This is crane C-14 being used to rebuild levees on April 14, 1922, following the flood that year.

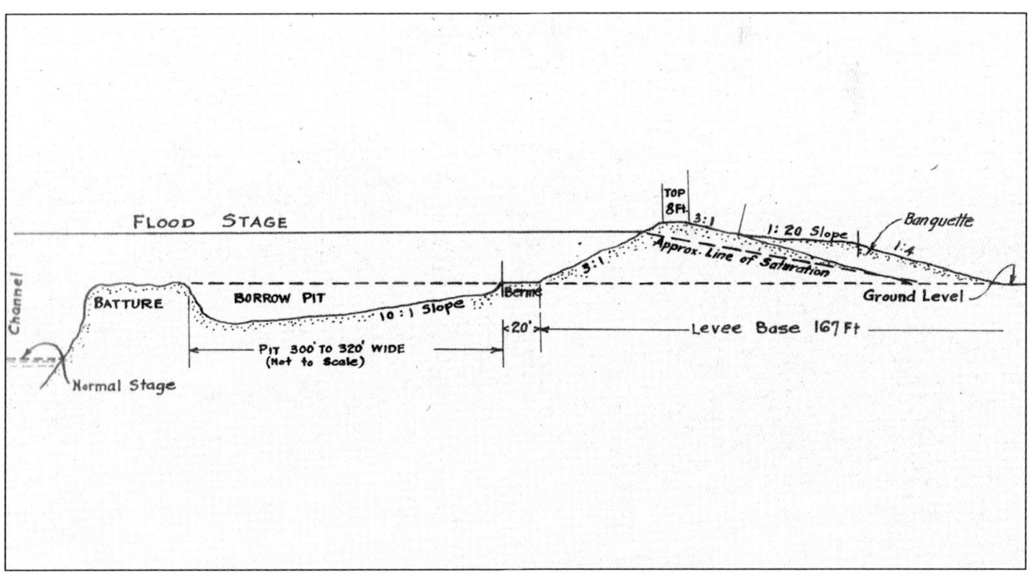

Although levee designs varied in height, grade, and density, they followed a similar concept. Teams obtained soil from a borrow pit in the batture—the area between the levee toe and the river. If the soil was swampy, it was often necessary to import borrow material. Engineers would strengthen weak levees with banquettes, berms, bank protection, or walls built into the crown.

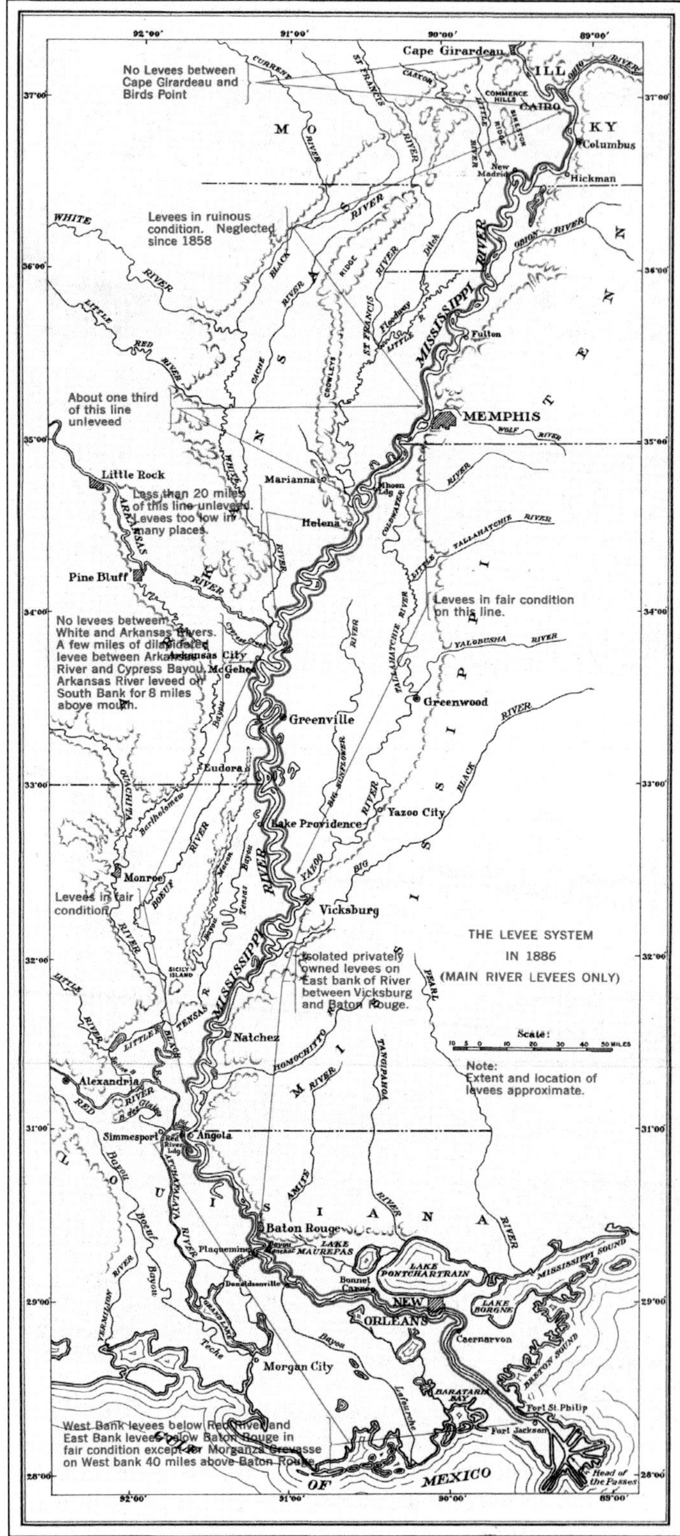

This map shows the status of levees in 1886. By that time, levees lined about two-thirds of the river, often only on one side, so there remained many gaps that left much of the valley unprotected. All it took was one gap or crevasse to flood huge stretches of the backcountry.

This 1965 photograph near Vicksburg, Mississippi, shows typical bank caving after being eroded by the current. In some places, erosion could result in the river meandering many thousands of feet from its original location.

To address bank caving, the commission installed revetment (from the French *revetir,* meaning "re-clothe"). Teams covered the banks with mattresses of willow branches buried along the bank. This photograph shows a revetment team in 1898.

Teams working on revetment would often stay in quarter boats, such as US quarterboat No. 237 (seen here), which docked near the worksite. The number of boats used depended on the size of the crew.

Prior to installing the revetment, it was necessary to grade the bank. This was most often a manual process involving workers with shovels or mules and plows. (Courtesy of the US Army Corps of Engineers, Memphis District.)

By the 20th century, the use of hydraulic methods (jets of water) proved faster and more effective than shovels and plows in grading the banks. The Corps also tried the same effect underwater for dredging, though less successfully due to the limited range of hoses.

Making the willow mattresses was also a time-consuming process and required extensive labor. The result was a mattress or frame laborers moved into place. (Courtesy of the US Army Corps of Engineers, Memphis District.)

Teams had to collect the willows, which grew primarily on islands or the shores of the Middle and Upper River, where sandy soil was abundant, and there was less frequent inundation.

View of mat at Helena, Ark.
Nov. 8, 1899.

Ships similar to the mattress-sinking steamship *Helena* (pictured) carried the willow downstream, where crews would construct mattresses, place and sink them, and cover them with earth and loose stones to keep them in place.

Once the mattresses were in place, work teams placed loose stones to hold them underwater. (Courtesy of the US Army Corps of Engineers, Memphis District.)

Because of the increasing scarcity of willow, the commission started experimenting with revetment made from wood, asphalt, or concrete in the 20th century. The commission installed this concrete slab weighing 192,000 pounds near Memphis in 1916.

The use of lumber mattresses remained popular in some locations as late as World War II, as with this mattress placed in 1944. Other materials used included brush, wire, and even bedsprings.

Placement of revetment took careful planning to affect the shape of the channel. This annotated aerial photograph shows where engineers planned revetment on Island 63 south of Helena, Arkansas, in 1973.

The commission also began to experiment with dredges to remove silt that collected in channels as sandbars, shallows, or widened banks. The first models were steam-powered bucket dredges, such as the dredge *Dupere*, seen here in 1924.

By the 1890s, the commission started to develop suction dredges, starting with the dredge *Alpha*. This undated photograph of *Delta*, the fourth-generation suction dredge, shows the enormous plant required at the time.

Dredge *Kappa*, the 11th-generation suction dredge, entered service in 1901 and remained in service until after 1927, the date of this photograph. It was a 199-foot side-wheel steamer with a 32-inch dredge pipe.

The Corps of Engineers continued snag removal using vessels little changed from Henry Shreve's original design. Seen here is the snag boat *Horatio G. Wright* of the Memphis District in operation in 1919.

Snag removal was also a labor-intensive operation, even with cranes and winches. (Courtesy of the US Army Corps of Engineers, Memphis District.)

A continual challenge on the river was the removal of sunken vessels blocking the channel. This photograph shows the steamboat *Chic* in 1894. It sank at Cottonwood Point near Caruthersville, Missouri.

Complicating river maintenance was a growing number of locally built canals and locks, which allowed access to the river from other bodies of water. Shown here is the 1921 construction of the Inner Harbor Navigation Canal (Industrial Canal), which connected the Mississippi River to Lake Pontchartrain, in New Orleans. (Courtesy of the US Army Corps of Engineers, New Orleans District.)

As a result of navigational improvements on the river made by the Mississippi River Commission and the Corps of Engineers, commerce boomed, especially in the South. Shown here is the steamer *Hallette* loading cotton in Vicksburg in the early 1900s.

As shown in this undated photograph of the steamer *Sprague*, towboats pushing long lines of barges became common by the 1890s. This saved money on fuel, which reduced the cost of goods shipped.

St. Louis grew to the fifth-largest city in the nation in the 1880s. This photograph shows the port of St. Louis in the 1890s. (Courtesy of the St. Louis Mercantile Library Association.)

This c. 1900 photograph shows the Memphis waterfront. Most vessels at the time remained rear-wheel steamers. (Courtesy of the US Army Corps of Engineers, Memphis District.)

The Mississippi River Commission operated several vessels, including the steamboat *Mississippi*, which the commission used to conduct annual inspection trips during high and low water. The current motor vessel *Mississippi* is the fifth of that name. The original was in service from 1882 to 1921 (Clay 30). (Courtesy of the US Army Corps of Engineers, Memphis District.)

The steamship *Inspector* was another Mississippi River Commission vessel, in use from 1915 to 1947. (Courtesy of the US Army Corps of Engineers, Memphis District.)

Seen here is the crew of the steamship *Inspector* around 1927 under Capt. Howard Fenton ("Inspector" 1). (Courtesy of the US Army Corps of Engineers, Memphis District.)

The major engineering challenge remained inundation. Particularly dangerous were crevasses, where water violently cut through levees. This photograph shows the Stella Crevasse near Yazoo, Mississippi, in 1891.

This photograph shows refugees in the flood of 1899. Refugees sought the closest higher ground, which often was on top of the levees.

Farmers dragged anything of value onto levees during floods, as shown in this group of refugees and farm animals near Hickman, Kentucky, in 1912.

Floods often lingered for weeks, as shown in this devastated town in Pointe Coupée Parish, Louisiana, in 1912. (Courtesy of the US Army Corps of Engineers, New Orleans District.)

The 1922 flood was a record-setting event. Although there were very few crevasses in the federal levees, they caused severe damage. This photograph shows attempts to repair Ashbrook Dike in Greenville Bends, Mississippi.

Teams repair a levee near Arkansas City after the 1922 flood. The flood generated widespread doubts about the efficacy of levees in protecting the Mississippi Valley.

Three

The Great Flood 1927

In 1927, the Mississippi River Commission consisted of, from left to right, (seated) Mississippi civil engineer Charles H. West, Col. Charles L. Potter, and Illinois civilian John Stipes; (standing) Col. George M. Hoffman, Col. Charles W. Kutz, Missouri civil engineer Edward Flad, Coast and Geodetic Survey member Capt. Robert L. Faris, and secretary Capt. Willis Teale. Potter had been commission president since 1920.

The 1927 flood resulted in a great number of crevasses in the Mississippi River levees (Manders, *Common Stock* 27–32). The first was on March 29, 1927, at Laconia, Arkansas. While crevasses

The second crevasse in the Mississippi River levees was on April 15, 1927, at Whitehall, Arkansas,

resulted in violent, crashing water, most floods came as gentle rises in water levels, as at this unknown location in Arkansas.

south of Memphis. One of the cities impacted was Helena, Arkansas, shown here.

With additional flooding on the White River, the flood extended into the Arkansas backcountry. Even cities as far from the river as Clarendon, Arkansas (seen here), received some flooding.

A major concern was the protection of Cairo (pictured), a growing industrial center at the time. A crevasse (April 16, 1927) at Dorena, Missouri, below the city, and several breaks in local levees upstream (April 17–30) greatly reduced flood stages at the city itself due to water diverted into the backcountry.

The next major crevasse in the federal levees occurred on April 20, at Knowlton, Arkansas. This led to further inundation of the backcountry, as at this unknown town in Arkansas.

One of the largest crevasses occurred on April 21, 1927, at Mound Landing, Mississippi, which produced a torrent of water. Within three days, this cut had flooded nearby Greenville to six feet.

With cities flooded for weeks on end during the 1927 flood, residents tried to make do, as with these raised walkways in Greenville, Mississippi.

The breach at Mound Landing widened to more than half a mile, flooding great swaths of Mississippi, including as far south as Vicksburg, shown here.

Even towns far inland flooded. One of these was Estill, Mississippi, a small stop about halfway between the Mississippi and Sunflower Rivers.

The wealthy were not immune from flooding; this photograph shows a flooded mansion in Mississippi. Many planters lost their entire crop that year, in addition to homes and outbuildings.

On the same day as the Mound Landing Crevasse, two breaks on Arkansas River levees at Pendleton and Medford, Arkansas, resulted in additional flooding. One of the cities impacted was Arkansas City, which most believed safe after the flooding of Greenville on the other side of the river.

Once flooded, cities remained underwater until the basin drained. This is how Arkansas City looked on May 6, 1927, two weeks after it flooded.

Pine Bluff, Arkansas, near Little Rock, also flooded after the crevasses on the Arkansas River on April 21, 1927.

On April 24, 1927, a ship knocked a hole in the levee at Junior Planation, Louisiana, about 45 miles downriver of New Orleans. This was the first crevasse in Louisiana.

Flood fighting was typically an all-hands-on-deck affair, but convicts and the poor and disenfranchised bore the brunt of the burden. Here, convicts are enlisted for flood fighting at a levee in Mississippi.

The usual method of rapidly raising levees was by building a wall of sandbags, as on this levee in Mississippi. The Corps of Engineers provided truckloads of canvas sandbags, which local laborers filled.

Another method, shown on the Torras, Louisiana, levees, was a flashboard—a plank reinforced by braces and sandbags—which could be thrown up quickly on the crown of a levee. (Courtesy of the US Army Corps of Engineers, New Orleans District.)

A mud box was a dirt- or sand-filled wooden structure added to the crown of a levee to increase its height. As with filling sandbags, usually poor or disenfranchised workers, such as the convicts shown here, did most of the work.

Paving levees, or the addition of concrete slabs, provided the best defense against temporary wave wash, as shown on these levees in Plaquemine, Louisiana. It was rare, however, due to the expense. (Courtesy of the US Army Corps of Engineers, New Orleans District.)

A faster method to protect against wave wash was to build a wooden wall on the waterline, braced against the crown. (Courtesy of the US Army Corps of Engineers, New Orleans District.)

In the past, people had sometimes destroyed a levee across the river to relieve flood heights in their own town. It was common for armed patrols to watch for saboteurs, though weak levees were more often the cause of crevasses. (Courtesy of the US Army Corps of Engineers, New Orleans District.)

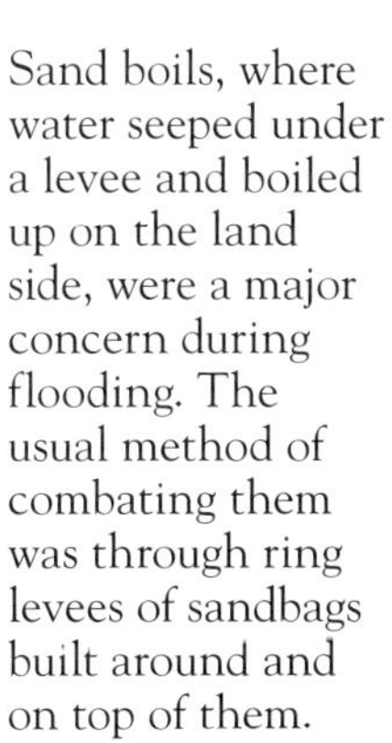

Sand boils, where water seeped under a levee and boiled up on the land side, were a major concern during flooding. The usual method of combating them was through ring levees of sandbags built around and on top of them.

After flood heights reached 21 feet in New Orleans, city leaders requested permission to make an artificial crevasse south of the city at Caernarvon, Louisiana (seen here), the site of a breach in 1922. This breach had lowered flood stages in the city. (Courtesy of the US Army Corps of Engineers, New Orleans District.)

Col. Charles Potter and the Mississippi River Commission approved the breach on the Caernarvon levee if Louisiana assumed responsibility for damages and obtained permission from the chief of engineers and secretary of war, who agreed to the move on April 26, 1927.

After evacuating residents to warehouses in New Orleans, Louisiana state engineer George C. Schoenberger oversaw the team that dynamited the levee at Caernarvon on April 29, 1927. (Courtesy of the US Army Corps of Engineers, New Orleans District.)

The Caernarvon levee in Louisiana eroded slowly. It took until May 1, 1927, for the crevasse to widen to 800 feet, followed by a noticeable drop in water levels at the levees in New Orleans. (Courtesy of the US Army Corps of Engineers, New Orleans District.)

By May 3, 1927, the gap in the Caernarvon levee reached 2,400 feet as the water pushed toward Breton Sound (Manders, *Common Stock* 38–41). (Courtesy of the US Army Corps of Engineers, New Orleans District.)

The next major crevasse occurred on May 3, 1927, at Cabin Teele, Louisiana, despite flood-fighting efforts. This was partly a result of the surging Yazoo River, which swelled from runoff from crevasses in Mississippi.

The crevasse at Cabin Teele quickly widened to more than 1,000 feet, inundating more of the Louisiana countryside.

Tallulah, Louisiana, on the edge of the Tensas River Basin, was the largest town flooded by the nearby Cabin Teele crevasse. It is less than 10 miles away.

Also flooded by the Cabin Teele crevasse was Monroe, Louisiana, west of Tallulah. Many people fled as far as Vicksburg (Daniel 45). (Courtesy of the US Army Corps of Engineers, New Orleans District.)

It took several weeks for the water to drain from Monroe, Louisiana—notice the change in water levels. (Courtesy of the US Army Corps of Engineers, New Orleans District.)

Among the many small communities flooded in Louisiana was Ashland. As with other locations, people and farm animals sought the high ground on the top of levees.

Additional crevasses at Glasscock, Mississippi, and Brabston and Bougere, Louisiana, on May 1, 1927, and at Winter Quarters, Louisiana, on May 4 flooded many farm regions north of Old River. (Courtesy of the US Army Corps of Engineers, New Orleans District.)

By mid-May 1927, the flood had descended down the Atchafalaya River. On May 11, the Bayou des Glaises levees—the strongest in the region—gave way. Pictured is a levee and flashboard at Simmesport, Louisiana. (Courtesy of the US Army Corps of Engineers, New Orleans District.)

From May 12 to May 16, 1927, there were additional breaches on the Bayou des Glaises levees on the Atchafalaya River. The crevasse at Moreauville, Louisiana, was one of the most dramatic, as water violently crashed through the levee. (Courtesy of the US Army Corps of Engineers, New Orleans District.)

The Atchafalaya River levee near Melville, Louisiana, breached on May 17, 1927, flooding more of backcountry Louisiana. (Courtesy of the US Army Corps of Engineers, New Orleans District.)

This photograph shows the flood levels in downtown Moreauville, Louisiana. While in some places (like the foreground), flooding was not very deep; in others, a boat was necessary to get to residences.

Residents of Melville, Louisiana, were wise enough to move their vehicles from the flooded area to higher ground. (Courtesy of the US Army Corps of Engineers, New Orleans District.)

The last crevasse occurred in the Atchafalaya River levees on May 24 near McRea, Louisiana. (Courtesy of the US Army Corps of Engineers, New Orleans District.)

The result of the Atchafalaya River breaches was the Louisiana backcountry being covered with water to a depth of several feet. The flood ruined crops, destroyed fences and property, and left behind a changed landscape.

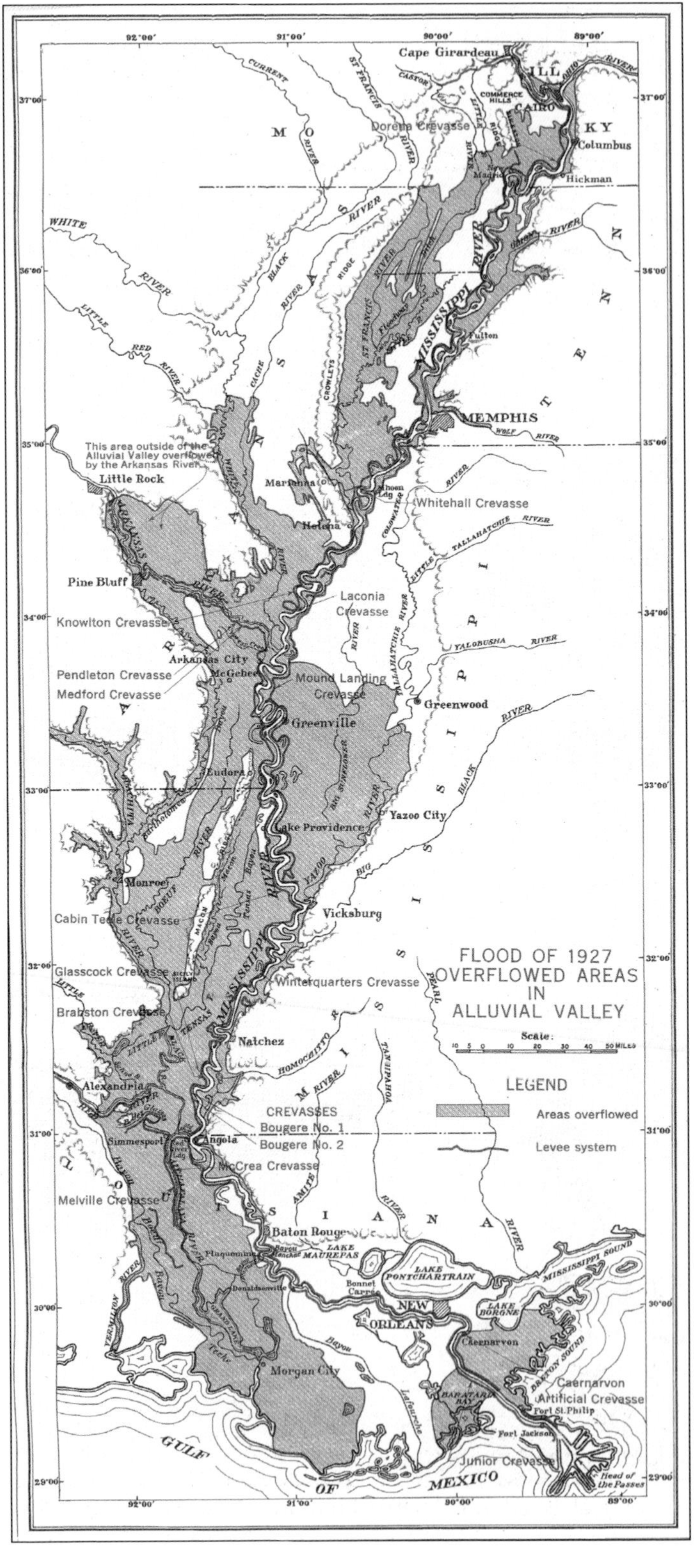

This map shows the regions inundated by the 1927 flood. It covers a huge area in seven states up to 70 miles wide and more than 600 miles in length.

Once storms were past, rescue efforts began. Although they could not pick up very many people, planes were instrumental in locating refugees. This plane is taking off from Natchez, Mississippi.

Here, a team takes a boat down Bayou Jack Road in Palmetto, Louisiana, to search for survivors. (Courtesy of the US Army Corps of Engineers, New Orleans District.)

As floodwaters rose, so did refugees to escape the flood, as in this house in Monroe, Louisiana. Some people even cut holes in their roofs to get out of their attics. (Courtesy of the US Army Corps of Engineers, New Orleans District.)

Levees were often the highest points in lowland communities, and many people gathered there to escape the flood. In Arkansas City, refugees set up tents on the levee because the fields nearby were still full of water.

People took shelter on levees, on bridges, or even in railcars, as seen here. Boats picked them up and deposited them at the nearest safe location. (Courtesy of the US Army Corps of Engineers, New Orleans District.)

Most refugees in flooded areas waited to be picked up and taken to camps, as with these men on Bayou Rouge near Palmetto, Louisiana. (Courtesy of the US Army Corps of Engineers, New Orleans District.)

For those whose homes were underwater, the 1927 flood was a terrible ordeal. For those on higher ground or who were wealthy enough to get supplies, it was like an extended holiday, as with these women in Mississippi.

Many towns remained flooded for weeks on end, requiring boats to get to stores or visit neighbors. (Courtesy of the US Army Corps of Engineers, New Orleans District.)

Federal leaders, including Secretary of Commerce Herbert Hoover (far left), Secretary of War Dwight Davis (with field glasses), and Red Cross vice chairman James L. Fieser (far right), helped direct the response to the flood from a boat on the Mississippi River. (Courtesy of the US Army Corps of Engineers, New Orleans District.)

Once rescued from flooded areas, people rode on barges to refugee camps near large towns, as with this group of refugees in Mississippi.

Refugees were housed in camps just outside the flood zone, like this one in Mansura, Louisiana. The Army donated surplus tents, cots, blankets, and other supplies.

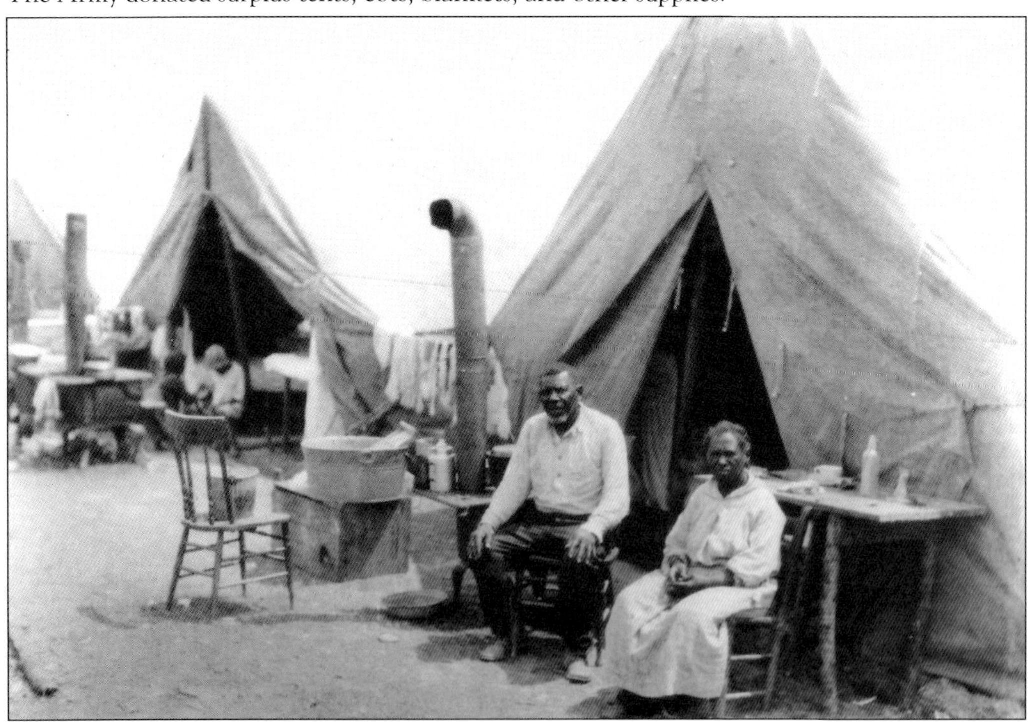

Many complained about the camps being beds of vice, while others complained about the mistreatment of African Americans. These complaints eventually resulted in an investigation by an advisory committee.

Some refugees, typically the disenfranchised, were forced to labor for low wages, but many others volunteered to protect their communities. (Courtesy of the US Army Corps of Engineers, New Orleans District.)

Whatever their circumstances, local workers played a critical role in repairing levees or filling sandbags, as these two men are doing. Raising levees required thousands of sandbags in a given reach.

The Army provided supplies, and the Red Cross managed the camps. As there was not a federal disaster agency at the time, the government relied mainly on charitable organizations and private donations or volunteers to meet the need.

The Red Cross ran canteens at the camps to provide coffee and other items to help improve refugee morale. The government worked directly with the Red Cross to operate the camps.

Among other services, the Red Cross provided medical care and inoculations against malaria, yellow fever, and other diseases, as with this woman in Mississippi.

In this photograph, volunteers feed refugees at a camp near Greenville, Mississippi. While the Army provided some field kitchens, it was mostly local churches and charities that fed refugees.

Once the flood was over, the primary focus of the Corps of Engineers and Mississippi River Commission was to repair federal levees, such as at Dorena, Missouri, seen here.

Although most water drained over time as the flood receded, local agencies had to pump out low-lying areas, as here in Mississippi. Some cities, such as New Orleans, had permanent pump stations.

The flood caused severe damage in many areas. Even houses that remained standing often had to be destroyed due to mold from long exposure to water. Notice the severe foundation damage to this house in Louisiana.

For the most part, private citizens were responsible for cleaning up their own property, as with this woman in Mississippi. There was no disaster assistance at the time other than that provided by churches and charities such as the Red Cross and Salvation Army.

The loss of livestock had a major impact on small tenant farmers, who sought to rescue animals by taking them to high ground such as levees.

In many places, the flood greatly changed the countryside, burying locations or carving out new lakes or holes. Here, a Mississippi man stands next to a nearly buried power pole. Local agencies had to pay for the restoration of such services without compensation.

This rail car crashed when a trestle near Vicksburg washed out. It is remarkable that there was not more loss of life from such accidents caused by the flood.

In some places, the flood shredded wooden structures or bent metal rails, as with this railroad in Louisiana. Even asphalt crumbled after long exposure to water.

Here, debris is trapped on the side of a bridge. Without federal funding, it took many years to clean up debris, rebuild houses and buildings, and restore services and facilities. Even so, some areas were permanently changed by the flood. (Courtesy of the US Army Corps of Engineers, New Orleans District.)

There was also a psychological impact of the flood. Look at the faces of these people in front of Sparta Ginn at Dockport, Louisiana (Daniel 180). Fear of flooding remained with those impacted for a generation.

Four

The Mississippi River and Tributaries Project

1928–1965

Chief of Engineers Maj. Gen. Edgar Jadwin, who had become nationally renowned for his role in responding to the flood of 1927, pushed through his own flood control plan for the Mississippi Valley, alienating many valley residents who opposed the necessary real estate concessions as land grabs.

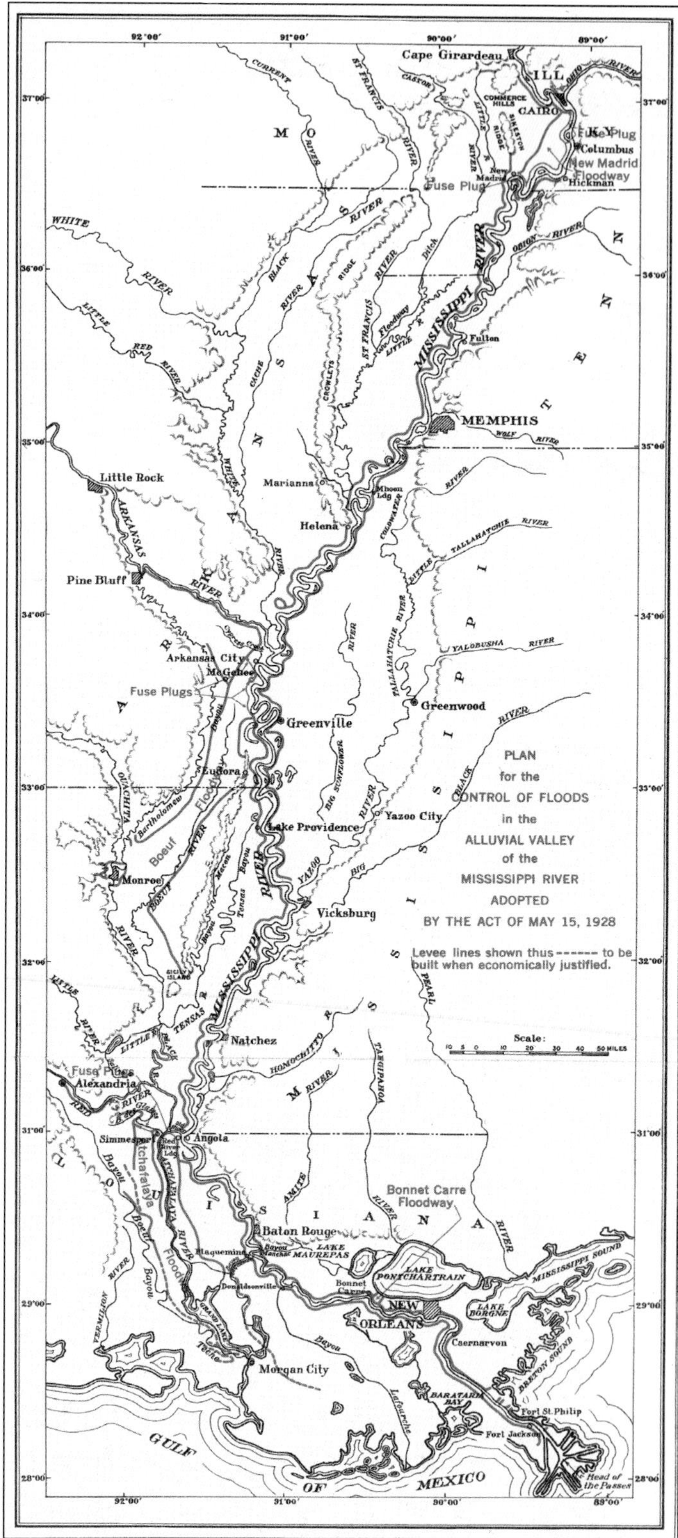

This map shows the major elements of the plan approved by the 1928 Flood Control Act, including the locations of the Birds Point–New Madrid Floodway, the Boeuf Floodway, the Atchafalaya Floodway, and the Bonnet Carré Spillway in Louisiana. These were designed to disperse floodwaters away from the Mississippi River to reduce flood stages at crucial population centers.

LEVEE CROSS SECTIONS

LOWER MISSISSIPPI RIVER

R.S.
L.S.
1937 H.W.
1927 H.W.
1922 H.W.
1882 H.W.
10′ CROWN
15′ ROADWAY
FEET
MEMPHIS GAGE
60
50
40
30
1:3.5
1:4
1:2
1:3
8′
8′
1:3
1:10
1:20
1:4
1:3.5
1:6.5
1:6
1882
1896
1914
1928
PRESENT

HORIZ. & VERT. SCALE IN FEET
0 20 40 60 80 100

CREST STAGES
MEMPHIS GAGE

HIGHWATER	1882-35.15
"	1922-42.5
"	1927-45.8
"	1937-50.4

DIMENSIONS OF CROSS SECTIONS

YEAR	HEIGHT FEET	BASE WIDTH FEET	AREA SQ. FEET
1882	9.0	53.0	274
1896	15.5	120.5	951
1914	24.0	200.0	2455
1928	27.0	260.0	3645
PRESENT	30.5	315.0	4956

PLATE III

The project also included dramatic changes in levee construction. This chart shows the development of levees from 1882 to 1937, which grew in height and width after each successive flood.

General Jadwin interpreted the 1928 act as placing the Mississippi River Commission under the US Army Corps of Engineers. The same year, he consolidated the commission under Brig. Gen. Thomas Jackson.

The 1928 act also created the US Waterways Experiment Station in Vicksburg to conduct scientific research on new construction. Based on the principle of similitude, hydraulic models simulated how water would act on a larger scale.

By 1929, construction of the Bonnet Carré Spillway was underway. The location chosen, about 30 miles upriver from New Orleans, had crevassed numerous times in past floods. (Courtesy of the US Army Corps of Engineers, New Orleans District.)

Construction of the Bonnet Carré Spillway included innovative concrete-laying techniques later used in better-known projects, such as the Hoover Dam. It was a 7,000-foot concrete weir with 350 bays that released floodwaters and thus lowered flood heights in New Orleans (Kemp 9-16). (Courtesy of the US Army Corps of Engineers, New Orleans District.)

Rather than using steel gates, the Bonnet Carré Spillway is unique in that it uses lower-cost wooden planks or "needles" to block the bays. A crane removes the needles one at a time, making opening the spillway a slow process. (Courtesy of the US Army Corps of Engineers, New Orleans District.)

This is an aerial photograph of the Bonnet Carré Spillway. Released floodwaters run just under six miles to empty into nearby Lake Pontchartrain. It is capable of releasing 250,000 cubic feet per second. (Courtesy of the US Army Corps of Engineers, New Orleans District.)

Maj. Gen. Lytle Brown, who became the chief of engineers after Jadwin in 1929, earned a reputation for being more considerate of complaints by Mississippi Valley residents in seeking new flood control solutions.

In 1929, General Brown reorganized Corps divisions into Upper and Lower Mississippi Valley Divisions and moved the Mississippi River Commission headquarters to Vicksburg, shown here in 1949.

In 1932, General Brown recommended the appointment of Brig. Gen. Harley B. Ferguson as the new president of the Mississippi River Commission. (Courtesy of the US Army Corps of Engineers, New Orleans District.)

The Mississippi River Commission in 1939 was the most brilliant in a generation. It included (from left to right) Brig. Gen. Harley B. Ferguson (president), US Coast and Geodetic Survey member Rear Adm. Leo O. Colbert, civilian Albert L. Culbertson of Illinois, civil engineer Edward Flad of Missouri, Col. Roger Powell, and civil engineer Harry N. Pharr of Arkansas.

Commission members during the early days of executing the project included some of the most talented engineers of the era. A good example is Harry N. Pharr, who formerly served as chief engineer of the St. Francis Levee Board and helped design many engineering features in that basin. (Courtesy of the US Army Corps of Engineers, Memphis District.)

In 1932, General Ferguson started a program of cutoffs—cutting off the bends in rivers—to lower flood heights by increasing the river velocity and preventing "stacking up" of floodwaters at bends. Here, Ferguson (second from left) and commission members watch the first cutoff at Diamond Point near Vicksburg.

To ensure a cutoff occurred gradually, Ferguson made a "pilot" cut, removed the final plug with dynamite, and allowed the cutoff to widen naturally. Seen here is a pilot cut being dredged in the Memphis District. (Courtesy of the US Army Corps of Engineers, Memphis District.)

The most winding bends on the Mississippi River were the Greenville Bends in Mississippi, where the river ran 100 miles over a 50-mile distance.

This aerial photograph shows the results of the Greenville cutoffs with annotations to show the new route of the river. Greenville, Mississippi, is to the right. Ashbrook cutoff is at upper left, Tarpley cutoff is at center, and Leland cutoff is at bottom.

After time, the river segments created by a cutoff often turned into oxbow lakes, such as Lake Ferguson, the remains of Bachelor Bend near Greenville, Mississippi.

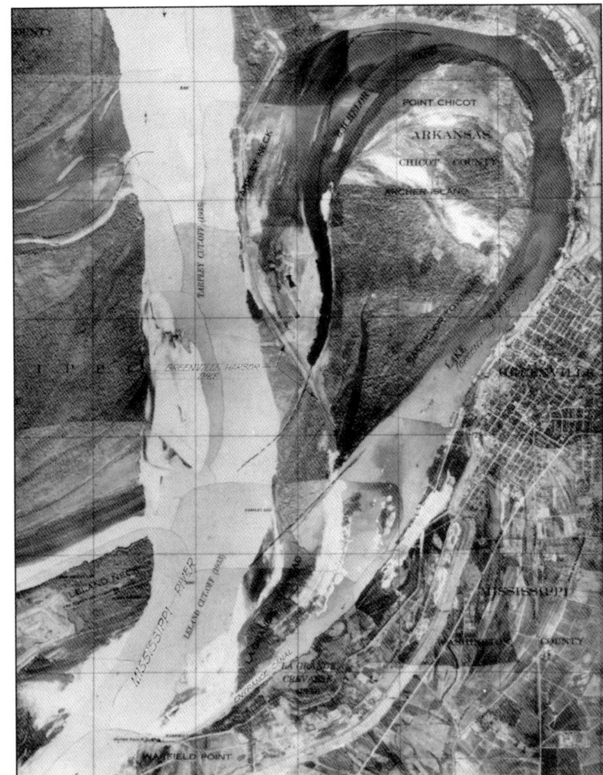

Dredging the river above and below the cut was a major part of General Ferguson's program, but he also used dredges to carve a shorter and deeper channel. This is the dredge *Pullen* in 1941 near Plaquemine.

After 1927, the Corps of Engineers introduced dustpan dredges like the dredge *Jadwin*. As shown here, a wide dredge pipe with a brush was ideal to remove sandy deposits from the bottom of a river.

Extensive revetment around new bends helped to reduce erosion after a cut. By the 1930s, the Corps of Engineers had almost exclusively adopted mattresses of articulated concrete blocks held together by steel cable. This photograph was taken near Vicksburg in 1935.

Using the mat-sinking plant, a barge designed for placement of revetment, workers would unroll the concrete mattresses on the shore at required locations. (Courtesy of the US Army Corps of Engineers, Memphis District.)

Laying the concrete revetment was a laborious task and required a large crew that traveled with the mat-sinking plant. (Courtesy of the US Army Corps of Engineers, Memphis District.)

The Atchafalaya Basin was a major distributary of the Mississippi River, but it was choked with swamps and debris. To make the floodway down the river efficient, engineers had to remove snags and dredge the Atchafalaya River and other waterways.

The 1937 flood in some ways proved a greater flood between Cairo and Memphis than even the 1927 flood. This photograph shows flooding around Cairo.

Work crews shored up levees around the clock at Ridgely, Tennessee, during the 1937 flood. Sandbags remained the first and fastest way to increase levee heights.

The Corps of Engineers put the Birds Point–New Madrid Floodway into operation for the first time in the 1937 flood. The fuse-plug levee was incomplete, forcing the Memphis District to dynamite the levee.

In 1939, Brig. Gen. Max C. Tyler became president of the Mississippi River Commission. He made three more cutoffs above Arkansas City. Including a natural cutoff at Yucatan, Mississippi, there were 16 cutoffs from 1929 to 1942, which shortened the river by roughly 500 miles.

Completed in 1952, the Mississippi Basin Model was the largest hydraulic model ever built and allowed hydraulic experiments on the entire Mississippi River and Tributaries Project at once. The Corps estimated it helped prevent more than $63 million in damages (Manders, *Common Stock* 75).

The Mississippi Basin Model's most innovative feature was its complex series of industrial controls to manage water flow. These controls replaced more than 250 laborers required to operate it manually.

Although the Mississippi River and Tributaries Project greatly reduced flooding in many locations, periodic flooding in riverside towns continued, as at Cape Girardeau, Missouri, in 1944.

In the flood of 1949, Wilkinson Point near Palmetto, Louisiana, was one of the locations that flooded, despite flood-fighting efforts. Here, bulldozers build roadways while a crane places stone and earth.

In Alton, Illinois, a man paints flood marks on the wall after the 1951 flood. In Illinois, 1844 was the flood of record. More recent floods had a lesser impact.

With the lowering of flood heights due to cutoffs, General Ferguson had recommended a smaller floodway at Morganza. Preliminary work on the Morganza Floodway started in the late 1930s. Here, crews complete pile tests.

This photograph shows the construction of the Morganza Floodway. It includes a 4,100-foot concrete spillway with 125 gates capable of diverting 600,000 cubic feet per second. It was completed in 1954 (Reuss 198-203).

This annotated aerial photograph shows the plan for the Morganza Floodway. It would come into use only after the inundation of lands set off by the guide levees. Overflow emptied into the Atchafalaya Basin. Melville, Louisiana, received increased flood protection due to changes in flow on the Atchafalaya River.

By 1950, studies showed that more than 25 percent of the Mississippi River was being diverted down the Atchafalaya Basin—up from less than 10 percent in the 19th century—and risked eventually capturing the river. To correct this issue, the Corps of Engineers started construction of the Old River Control Structure in 1955 (Reuss 210-242).

This aerial photograph shows the entire Old River Control Structure. North of the Low Sill Structure was the Overbank Structure, which allowed increased overflow during flood conditions. A lock and dam to the south bypassed the structure, allowing traffic via Old River to the Atchafalaya River.

Completed in 1963, the Low Sill Structure (seen here) maintained a 30 percent diversion from the Mississippi to the Atchafalaya River in normal conditions.

The Corps of Engineers also built 26 locks and dams on the Upper River to maintain a nine-foot channel. Construction on the first dams started not long after their 1930 authorization and continued into the 1940s. Lock and Dam 14 (pictured here) was just north of Davenport, Iowa.

One of the earliest locks completed was Lock and Dam 26, north of St. Louis. Work started in 1934 and was completed in 1938.

The final addition to the Mississippi River locks and dams addressed navigation around the Chain of Rocks, a seven-mile stretch of rapids near St. Louis, where depths sometimes were as low as five feet.

Added officially as Lock and Dam 27, the Chain of Rocks lock and dam included a bypass canal and two locks. Work started on the canal in 1945, and the canal and locks were complete in 1953.

After 1955, the Mississippi River Commission completed a study that resulted in the first major changes in the flood control plan since 1928. Shown are the commission members in 1955; from left to right are Brig. Gen. William E. Potter, Missouri civil engineer E.F. Salisbury, Adm. Leo Colbert, Brig. Gen. John R. Hardin (president), Louisiana civil engineer DeWitt L. Pyburn, Egbert A. Smith of Illinois, and Col. John L. Person.

Construction of dikes remained important, though most were now made of stone. This is the Victoria Bend dike north of Rosedale, Mississippi.

Five

Challenges to the Project 1965–2011

The Gulf Intracoastal Waterway provided an inland waterway from Brownville, Texas, to Apalachee Bay, Florida. The Corps of Engineers built it piecemeal as approved by Congress through the 1930s. (Courtesy of the US Army Corps of Engineers, New Orleans District.)

The Inner Harbor Navigation Canal from the Mississippi River to Lake Pontchartrain remained one of the highest-traffic locks through New Orleans. (Courtesy of the US Army Corps of Engineers, New Orleans District.)

Barge traffic was much more fuel-efficient and environmentally friendly than shipping by rail or truck since a single tug could push numerous barges of goods. In 1981, the *Miss Kae-D* set a record pushing 72 barges near Memphis.

In addition to floods, a major threat to the river was hurricanes, which caused severe erosion. Seen here are before and after photographs of Grand Isle, Louisiana, after Hurricane Betsy directly hit New Orleans in 1965. (Courtesy of the US Army Corps of Engineers, New Orleans District.)

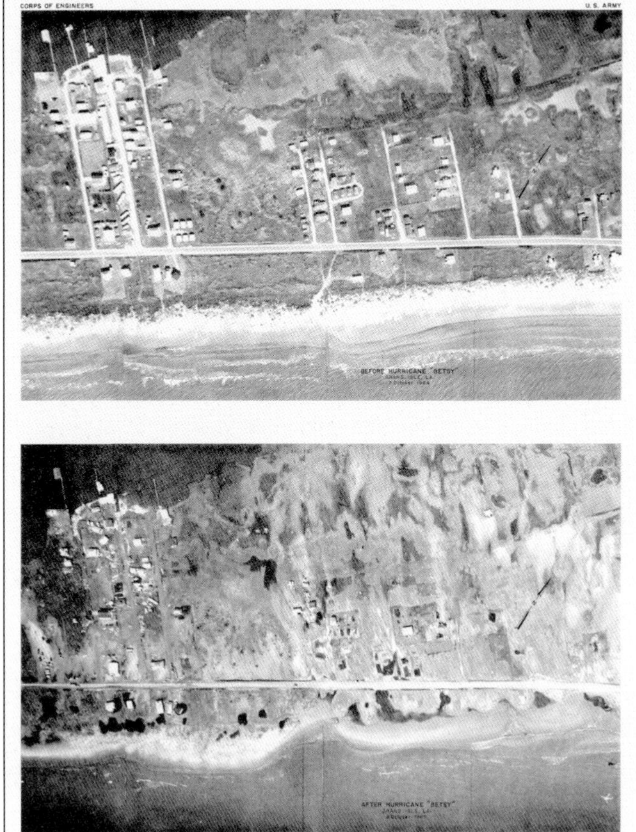

Hurricane Betsy flooded many parts of New Orleans, including Arabi. (Courtesy of the US Army Corps of Engineers, New Orleans District.)

This photograph shows flooding in the Lower Ninth Ward of New Orleans after Hurricane Betsy. The shut-down Inner Harbor Navigation Canal is in the foreground. (Courtesy of the US Army Corps of Engineers, New Orleans District.)

After more than a decade with little or no flooding on the Mississippi River, the 1973 flood broke numerous flood records. This photograph shows inundation in Alton.

Across the river from Alton, Crystal City, Missouri, also received severe flooding in 1973. Crevasses and overtopping throughout the basin resulted in the inundation of thousands of square miles.

Volunteers at Choteau Island, Missouri, place sandbags during the 1973 flood fight. There was severe flooding in Missouri and Illinois due mainly to local levee failures.

Polyethylene plastic sheeting came into widespread use in flood fighting for the first time during the 1973 flood, as seen here at Tiptonville, Tennessee.

An Army truck picks up refugees in Vicksburg during the 1973 flood. Even where the flood did not break flood stage records, in many locations, it did break duration records.

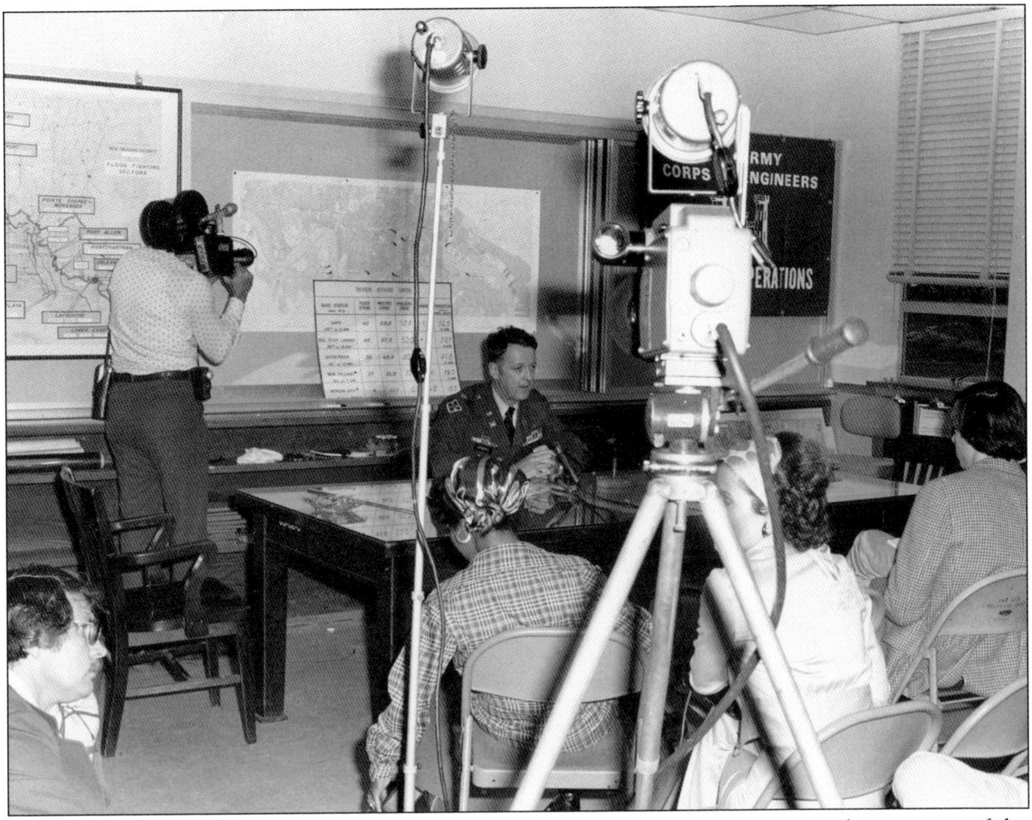

Then–New Orleans District commander Col. Elvin R. Heiberg III announces the opening of the Bonnet Carré Spillway near New Orleans in 1973. He went on to become the chief of engineers from 1984 to 1988.

A crane opens the Bonnet Carré Spillway one needle and bay at a time during the 1973 flood. This is a unique design that requires a gradual opening of the spillway.

The Bonnet Carré Spillway was fully open by April 7, 1973. It can divert up to 250,000 cubic feet per second, or more than 1.8 million gallons per second.

The Bonnet Carré Spillway remained fully open for a record 75 days in 1973 ("Historic Operations" 1). The spillway comes into use most often of any floodways on the Mississippi River—about once a decade on average—but often only at limited capacity or for a brief time.

At the Old River Control Structure, a boat hit the Low Sill Structure in 1973, causing minor damage. The Corps would later add a barrier to keep barges from being sucked into the intake.

The 1973 flood scoured a hole in the foundation of the Low Sill Structure the size of a football stadium and destroyed a guide wall (Reuss 243-245). This photograph shows the missing guide wall, similar to the one on the right side of the photograph.

Emergency repairs to the Old River Control Structure after the 1973 flood included filling the scour hole in the Low Sill Structure with grout and building a rock dike to replace the guide wall.

To reduce risk at Old River by lowering stages, the Corps of Engineers placed the Morganza Floodway into operation for the first time in 1973.

After the 1973 flood, the Corps of Engineers built the Auxiliary Structure at Old River to supplement the Low Sill Structure. Construction started in 1981.

Completed in 1986, the Auxiliary Structure at Old River provided additional overflow capacity during floods to assist the Low Sill Structure in maintaining flow down the Atchafalaya River.

This aerial photograph shows all of the Old River components: the Auxiliary Structure and Low Sill Structure at bottom left and right, the Overbank Structure at far right, and the lock and dam at upper left.

While the Corps of Engineers built and operates numerous navigation and flood risk reduction structures, most incorporate large areas of greenways and bodies of water widely used for recreation, which became increasingly popular after the environmental movement of the 1960s.

While the Corps of Engineers enforces the Endangered Species Act on its properties, it has long studied the impact of wildlife on its projects, including invasive species such as the zebra mussel.

In the 1970s, Louisiana State University professor Sherwood Gagliano first documented the extent of coastal erosion and wetland loss in Louisiana. This produced efforts to reduce the loss through various means. (Courtesy of the US Army Corps of Engineers, New Orleans District.)

By the 1960s, the Corps of Engineers began researching the beneficial uses of dredged material. By placing rich river sediment in targeted locations, it could restore and expand wetland areas.

One of the key locations to rebuild wetlands was the Atchafalaya Basin, which had been decimated by earlier efforts to clear out a floodway for the Mississippi. Levees sometimes prevented the flow of water needed to maintain wetlands. (Courtesy of the US Army Corps of Engineers, New Orleans District.)

Col. Thomas Sands, who was New Orleans District commander (1978–1981) and Mississippi River Commission president (1984–1989), helped to push the development of a plan to save the Atchafalaya Basin (Reuss 300). (Courtesy of the US Army Corps of Engineers, New Orleans District.)

B.E.M. Skerritt, president of the Greater Atchafalaya Basin Council, cooperated with the Corps of Engineers to formulate a plan to save the Atchafalaya Basin, and promoted it locally (Reuss 253). (Courtesy of the US Army Corps of Engineers, New Orleans District.)

In 1981, Louisiana governor David Treen shows the Atchafalaya Basin plan to Pres. Ronald Reagan. The plan divided the basin into management units to develop wetlands and greenways. It became a model for later wetland restoration efforts. (Courtesy of the US Army Corps of Engineers, New Orleans District.)

One method of reducing wetland loss was the diversion of Mississippi River water and sediment to coastal wetlands in an attempt to restore the natural rhythms of the river. Built in 1991 about 15 miles downriver from New Orleans, the Caernarvon Freshwater Diversion Structure diverts about 8,000 cubic feet per second into the Caernarvon Delta to restore wetlands. (Courtesy of the US Army Corps of Engineers, New Orleans District.)

Completed in 2002, the Davis Pond Diversion Structure, about 15 miles upriver from New Orleans, can divert more than 10,000 cubic feet per second. (Courtesy of the US Army Corps of Engineers, New Orleans District.)

Hurricane Katrina in 2005 had a tremendous impact on the Mississippi River. Many levees, such as this one in Louisiana, received damage, requiring a lengthy rebuilding project.

Brig. Gen. Robert Crear was the commander of the Mississippi Valley Division at the time of Hurricane Katrina, and it fell to him to organize the response.

A major problem after Hurricane Katrina was a series of breaches in drainage canals near Lake Pontchartrain. Using helicopters, the Corps of Engineers dropped giant debris bags filled with sand until the holes were blocked.

Once the Corps of Engineers stopped water flowing into the city, the priority became "unwatering" the city. Engineers from around the world descended on New Orleans with portable pumps, but major progress waited on the repair of the permanent pump stations.

The 2011 flood was the most severe flood since 1927. It was also the first time the Mississippi River and Tributaries Project was able to pass a flood without major federal levee breaches. For the second time, the Corps put the Birds Point–New Madrid Floodway into operation by exploding the fuse-plug levee.

In this photograph, increased flow through the Old River Control Structure helps to maintain flow from the Mississippi during the flood of 2011. It was the greatest test of the structure after the damage that occurred in the 1973 flood.

The 2011 flood marked the second time the Morganza Floodway went into operation. The use of the floodway demonstrates that each component of the system helps to protect the Mississippi Valley.

Bibliography

Camillo, Charles A., and Matthew T. Pearcy. *Upon Their Shoulders: A History of the Mississippi River Commission from its Inception through the Advent of the Modern Mississippi River and Tributaries Project*. Vicksburg, MS: Mississippi River Commission, 2004.

Clay, Floyd M. *A Century on the Mississippi: A History of the Memphis District, US Army Corps of Engineers, 1876–1981*. Memphis, TN: US Army Corps of Engineers, Memphis District, 1986.

Daniel, Pete. *Deep'n as It Come: The 1927 Mississippi River Flood*. Fayetteville, AR: University of Arkansas Press, 1996.

"Historic Operations of the Bonnet Carré Spillway." www.mvn.usace.army.mil/Missions/ Mississippi-River-Flood-Control/Bonnet-Carre-Spillway-Overview/Historic-Operation-of-Bonnet-Carre/. Accessed May 8, 2021.

"Inspector (Towboat, 1915-1947)." digital.library.wisc.edu/1711.dl/N2UG7Q7WGZN3N87. Accessed May 1, 2021.

Kemp, Emory L. *Stemming the Tide: Design and Operation of the Bonnet Carré Spillway*. Chicago, IL: Public Works Historical Society, 1990.

Manders, Damon. *The Cutoff Plan: How a Bold Engineering Plan Broke with US Army Corps of Engineers Policy and Saved the Mississippi Valley*. New York, NY: Nova, 2016.

———. *Improving the Common Stock of Knowledge: Research and Development in the US Army Corps of Engineers*. Vicksburg, MS: US Army Research and Development Center, 2009.

Manders, Damon, and Brian Rentfro. *Engineers Far from Ordinary: The US Army Corps of Engineers in St. Louis*. St. Louis, MO: US Army Corps of Engineers, St. Louis District, 2011.

Reuss, Martin. *Designing the Bayous: The Control of Water in the Atchafalaya Basin, 1800–1995*. Alexandria, VA: Office of History, US Army Corps of Engineers, 1998.

Consistent with our mission to preserve history on a local level, this book was printed in South Carolina on American-made paper and manufactured entirely in the United States. Products carrying the accredited Forest Stewardship Council (FSC) label are printed on 100 percent FSC-certified paper.